# Triddlers

## Color

### Volume 1

Smart Things Begin With Griddlers.net
Copyright © All Rights Reserved. www.griddlers.net

Author: Griddlers Team
Compiler: Rastislav Rehák
Cover design: Elad Maor
Editor: Hagit Maor

Contributors: 3omy3, Agrippina, allu78, amsterdam, baba44713, bohus, Breiprei, carootje, cf12, cherith, crevette, daelyn, DevilsladyP, dianaraven, douglas_itoh, Eilleen, Erlik, esra, fineke, FlutePicc, fockii, Inbar_b_y, ivanrehak, JadeyBaby, jamis, JynxsMom, Layla, Lui, Lyska, Makarios, myvalice, oldyard, sablett, spisula, StoVoKor5506, talroni, tuckers, ulka, vixen999, wiggles, yolinde

ISBN: 978-9657679319

Griddlers - Triddlers: Color, Volume 1. Copyright © Griddlers.net. All rights reserved to A.A.H.R. Offset Maor Ltd – (operating www.griddlers.net). Printed in the US. No part of this book may be used or reproduced in any manner whatsoever without written permission except in the case of brief quotations embodied in critical articles or reviews.

For more information:
Email: team@griddlers.net
Website: http://www.griddlers.net

# Triddlers Rules and Examples

Triddlers are picture logic puzzles that use number clues around a grid to create an image.

The clues encircle the entire grid in three directions.

---

Each clue indicates a group of contiguous triangles of like color.

---

Between each group there is at least one empty triangle.

---

The clues are already in the correct sequence.

---

Groups of different colors may or may not have empty triangles between them.

---

In Black and White puzzles the clues are always black and the background is always white.

In Color puzzles the background may or may not be white.

The colored shape on the puzzle top-left corner indicates the color of the background.

# Solving a Puzzle

The gray arrow marks the direction of the clues as well as the direction they should be placed on the grid.

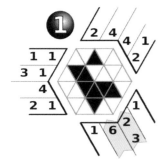

If we perform overlapping counting and count from the bottom up and from the top down, we can place 5 triangles of clue **6** on the grid. We can do the same for clues (**2,3**).

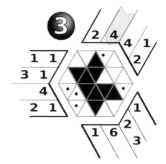

There are already 1 triangle and 2 triangles in the line of clue **4**.
We will add a triangle between them to make it a group of 4.

Clue **4** already has 4 triangles on the grid. We can mark the rest as background color.

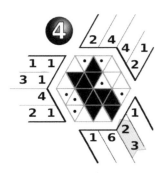

The triangle between two blocks has to be marked as background. Now we can complete the line of clues (**2,3**).

We can add the 4th triangle needed for clue 4 and complete the line of clues (**1,2**).

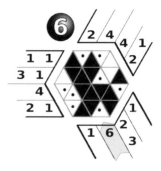

There is only one empty triangle left to complete clues (**1,1**) and finish the puzzle.

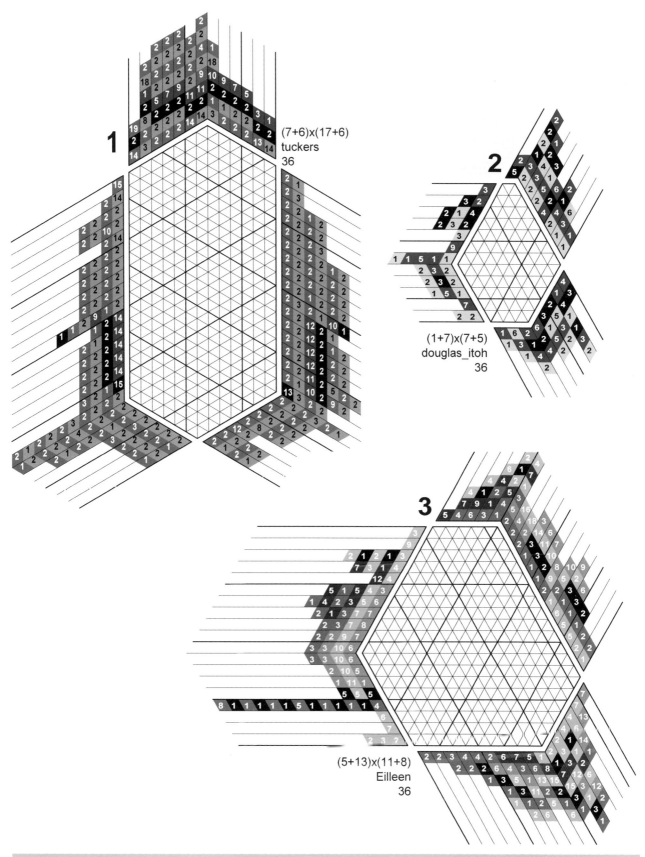

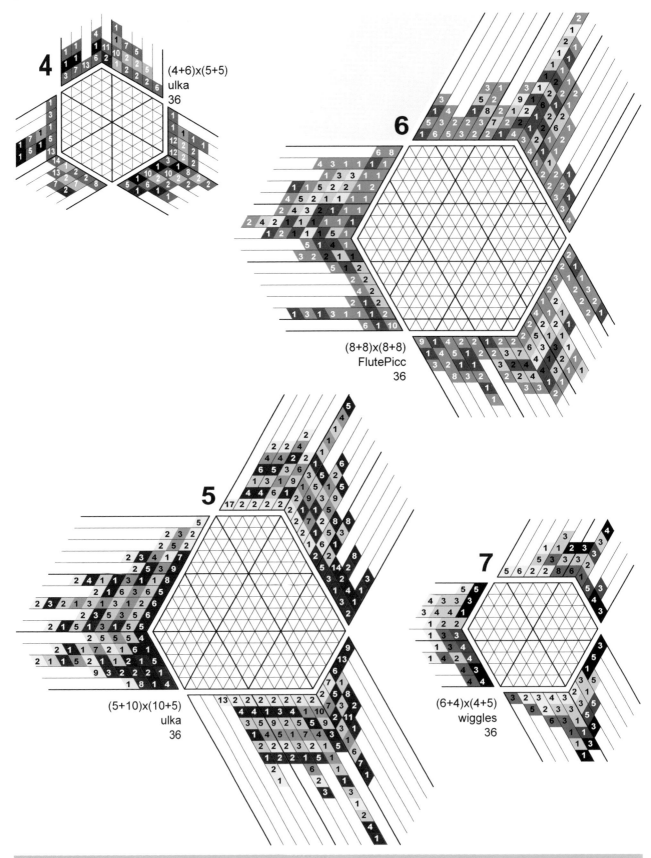

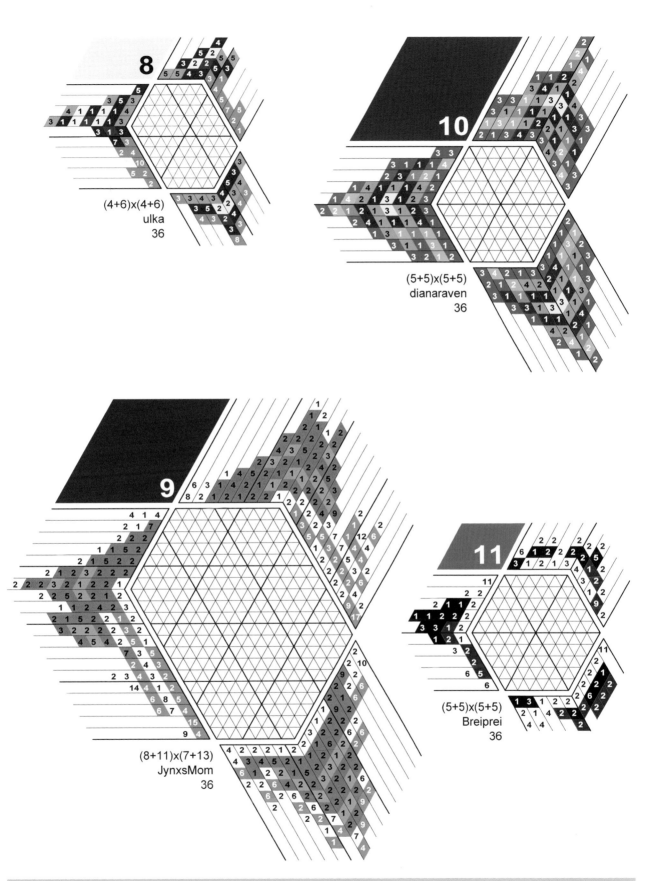

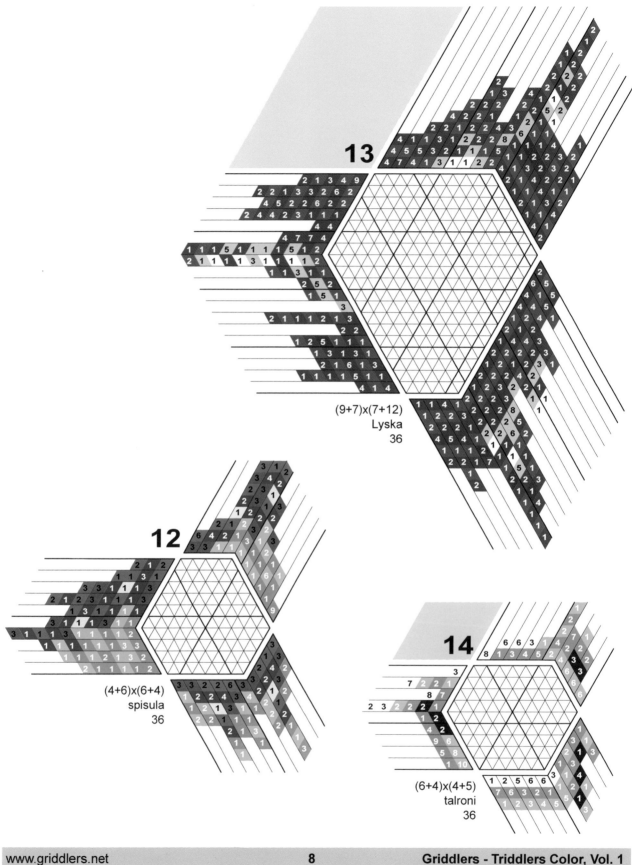

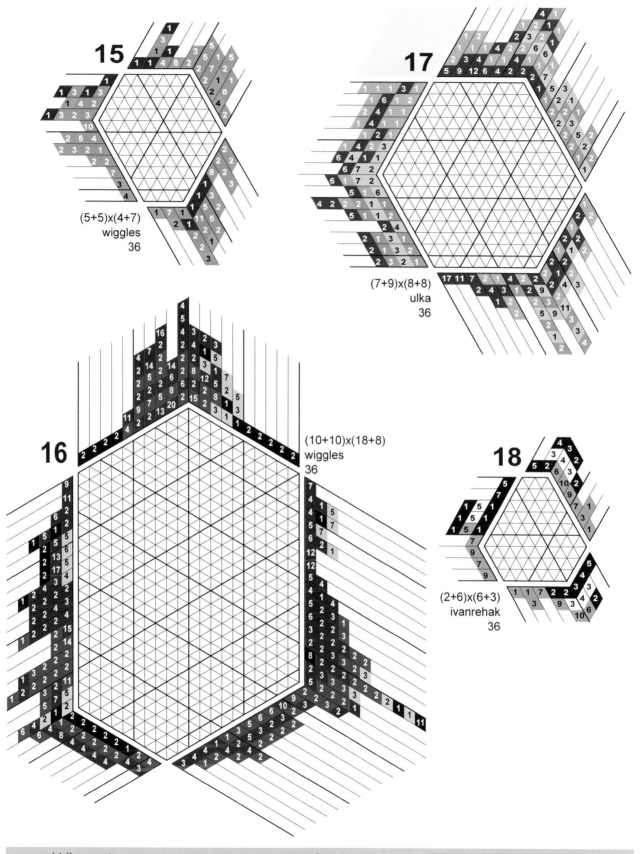

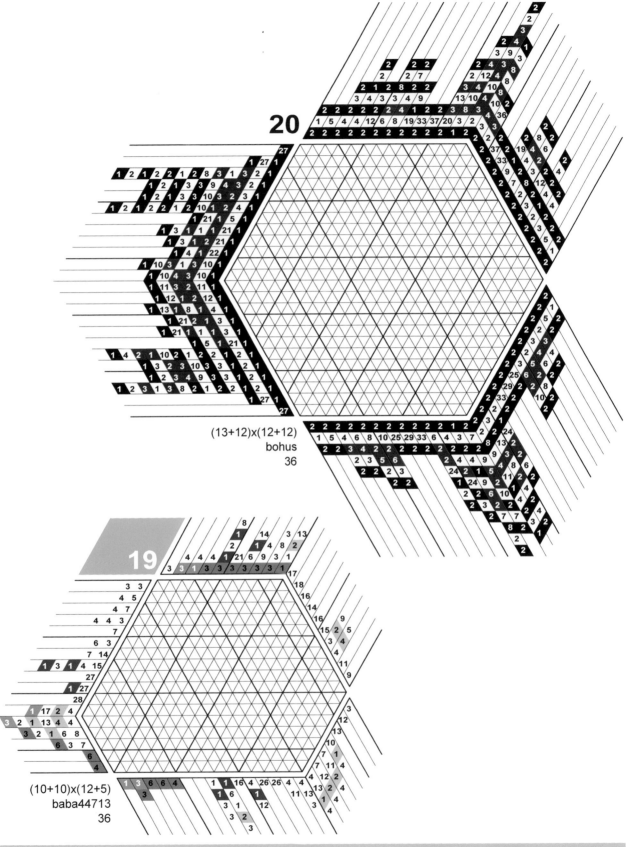

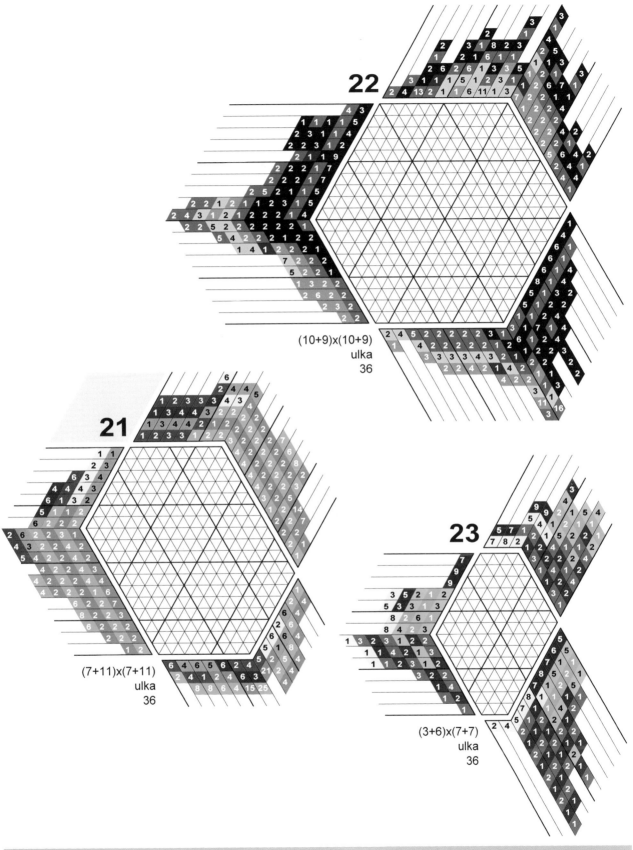

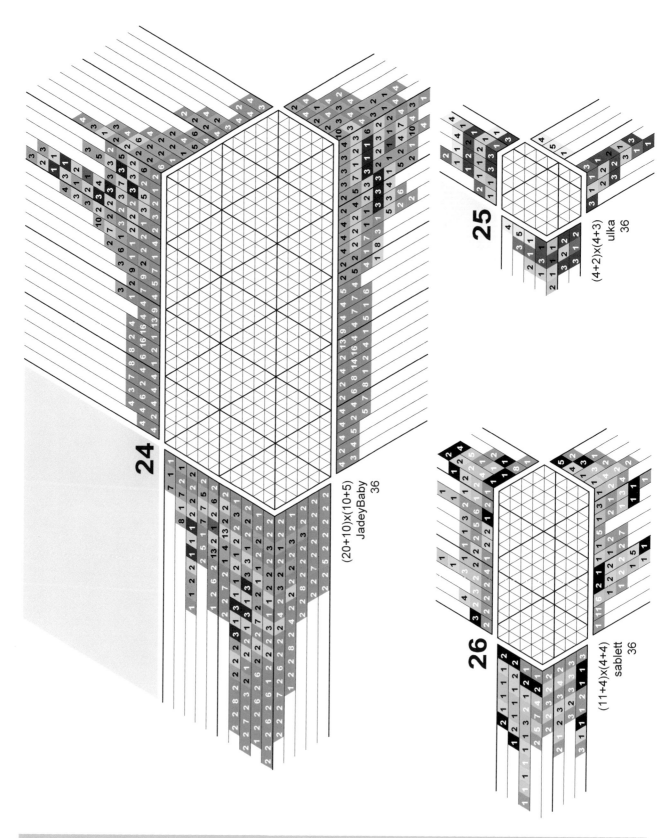

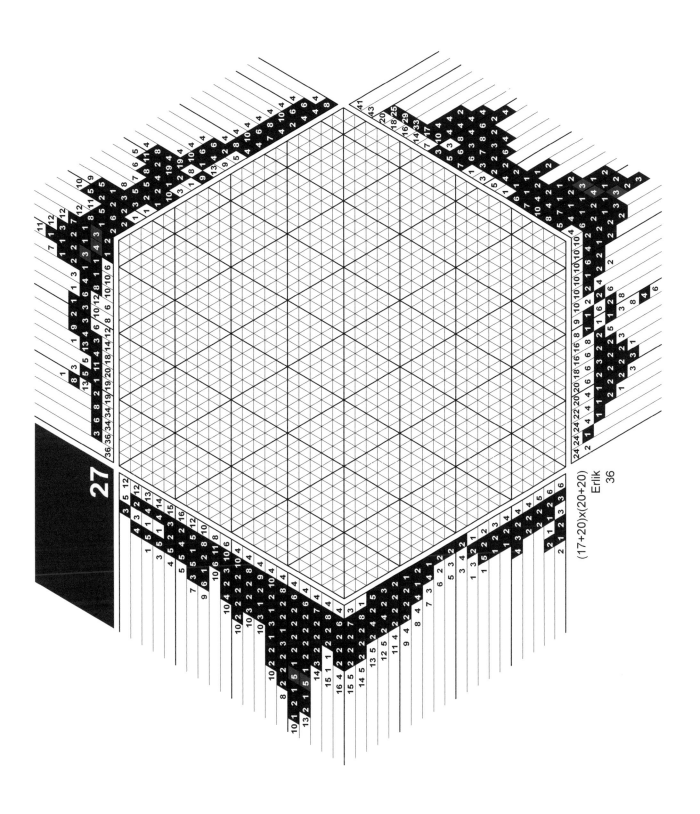

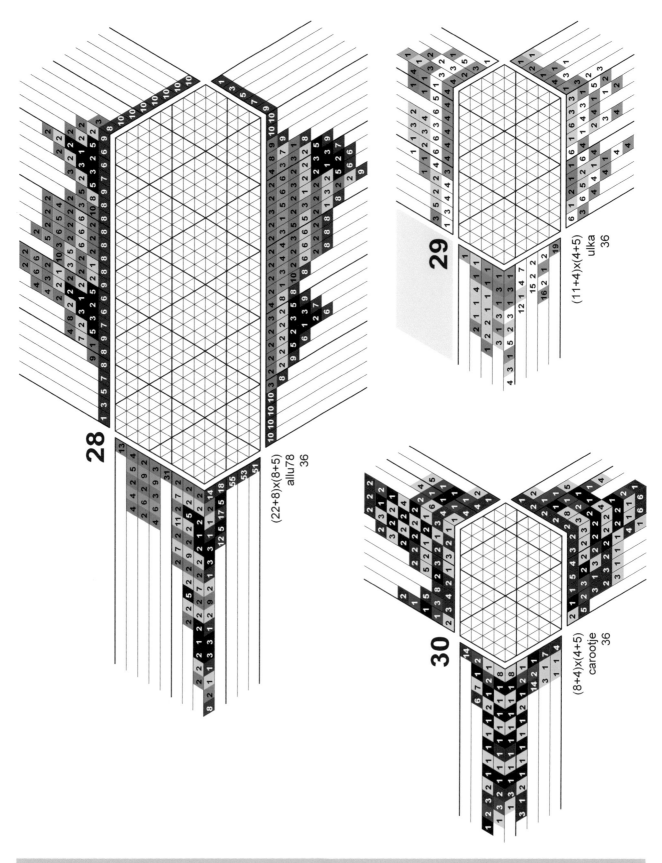

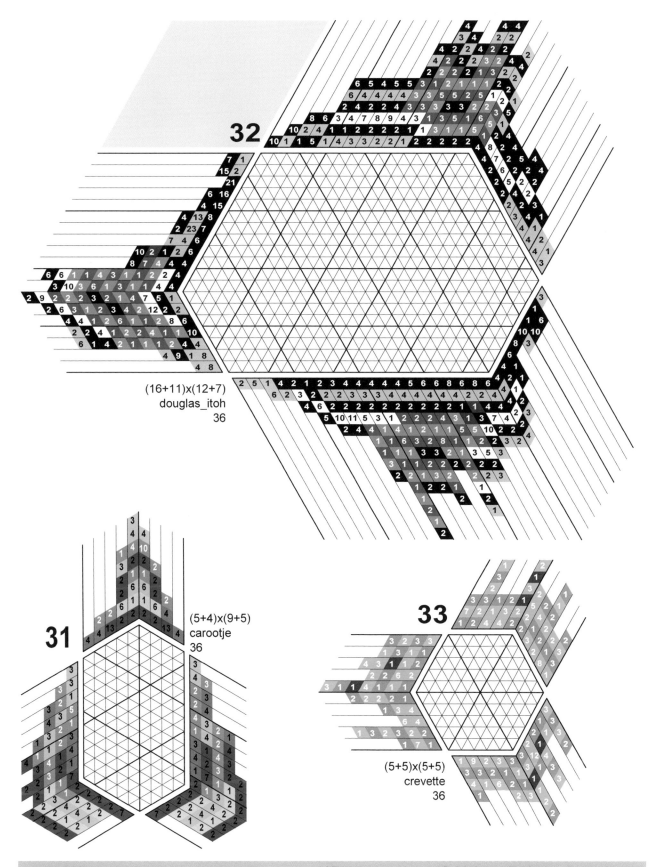

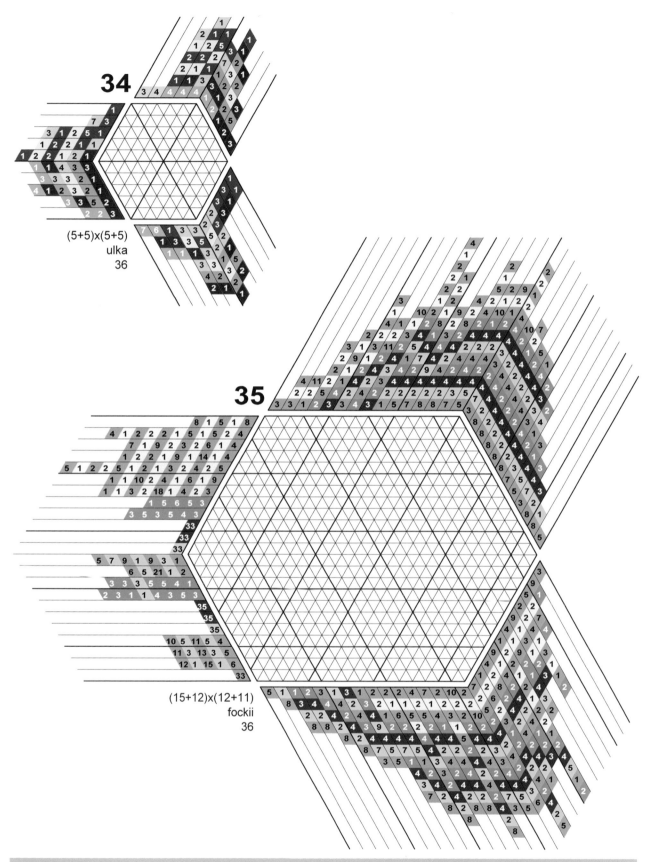

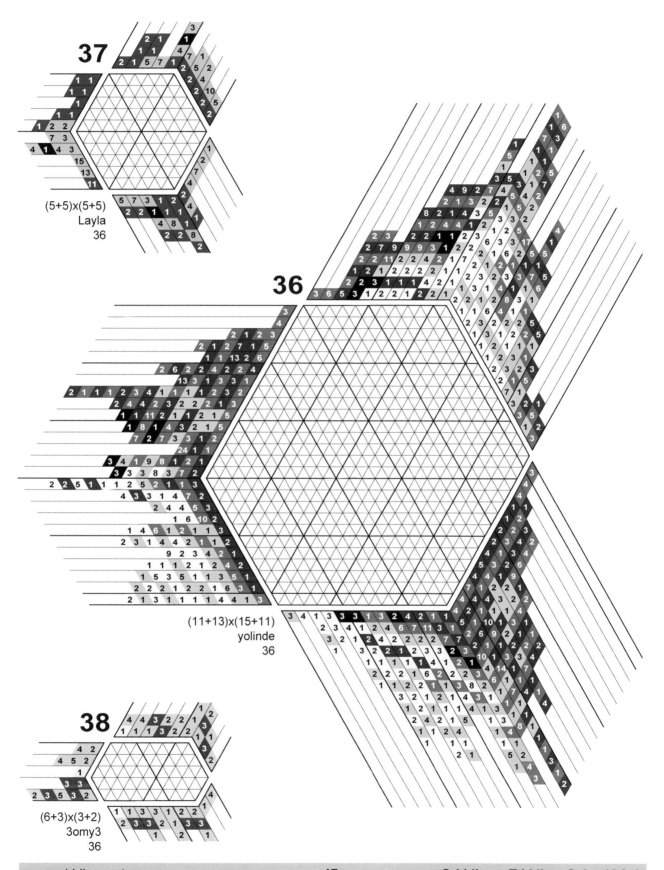

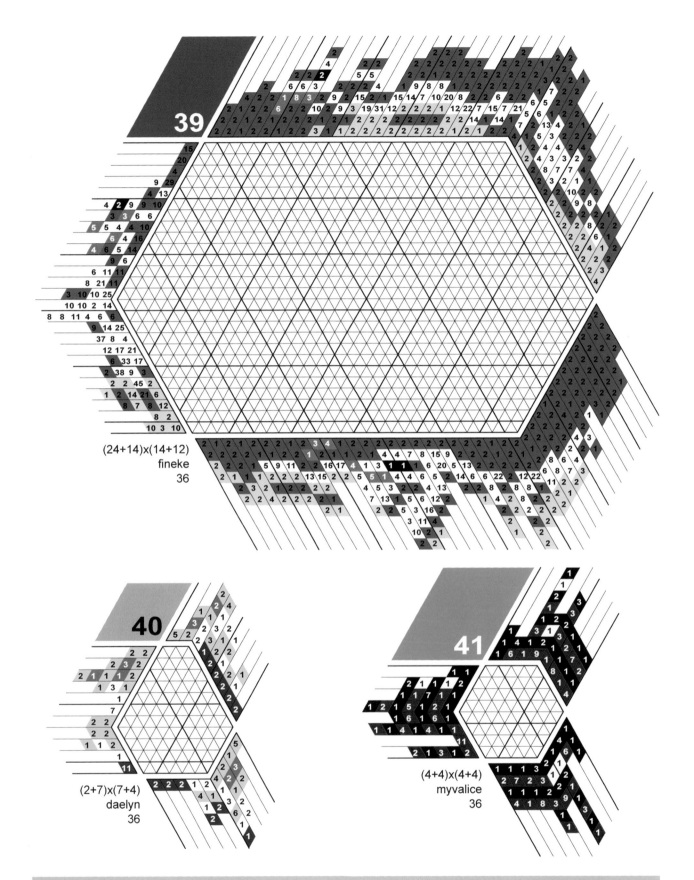

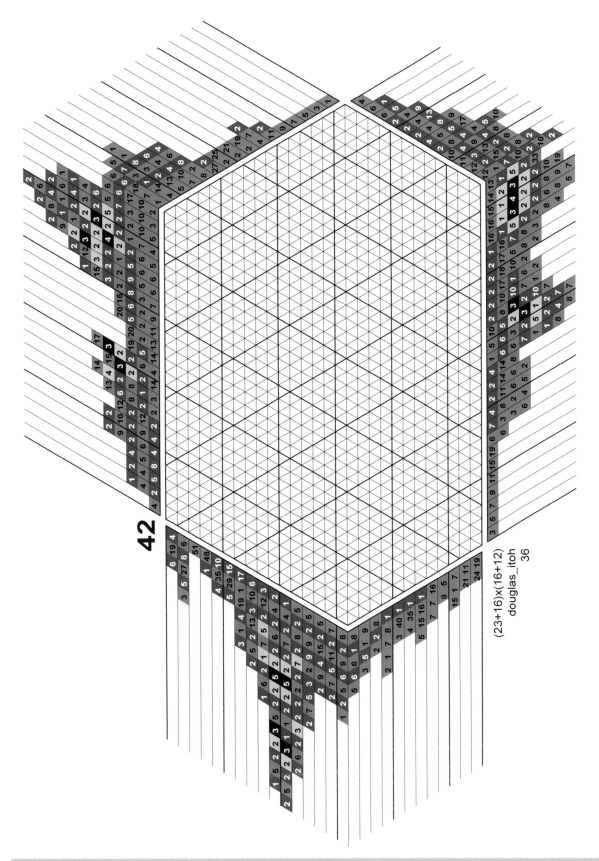

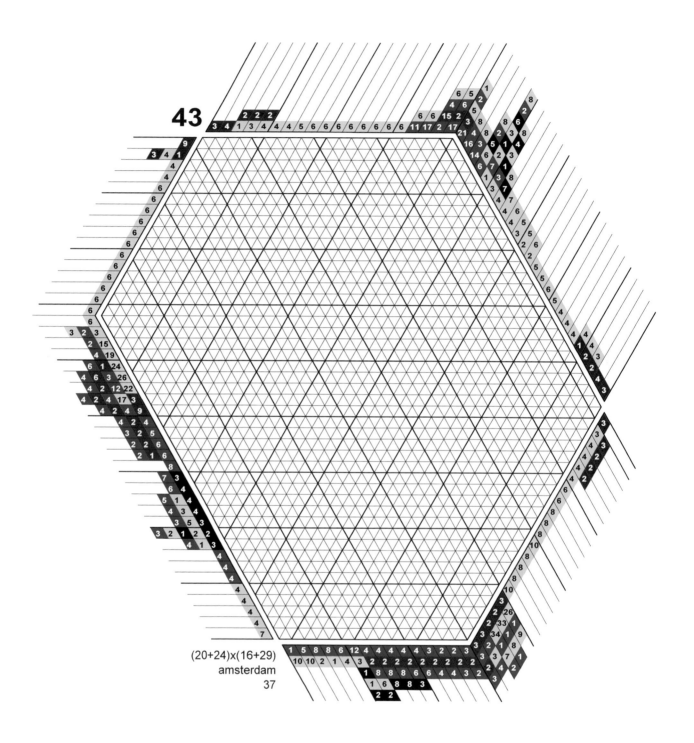

**43**

(20+24)x(16+29)
amsterdam
37

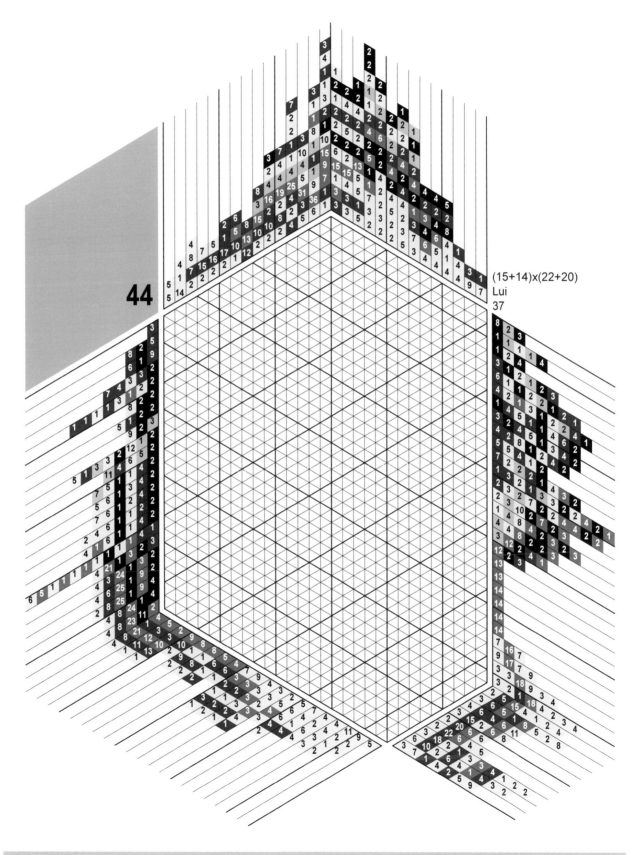

(15+14)x(22+20)
Lui
37

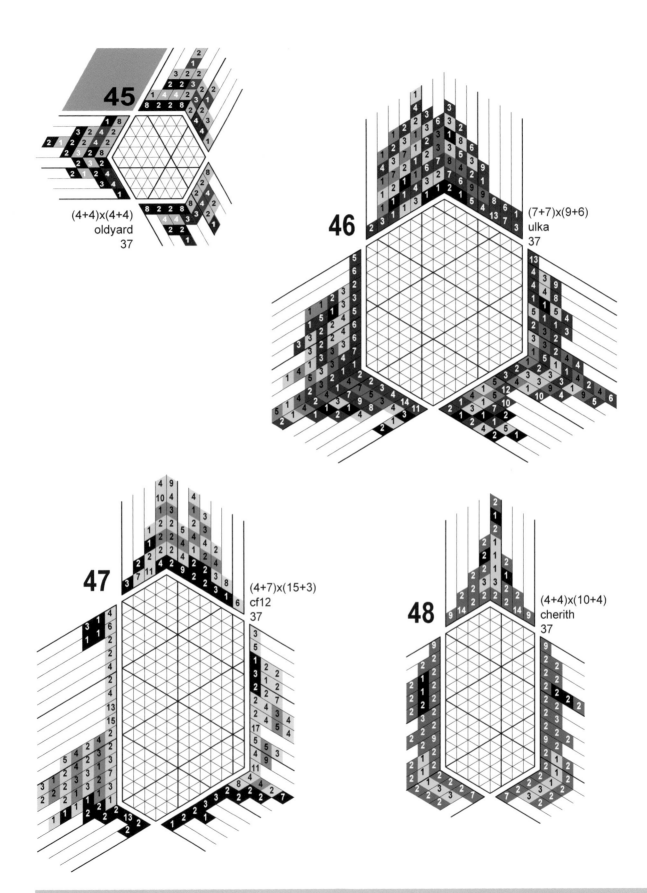

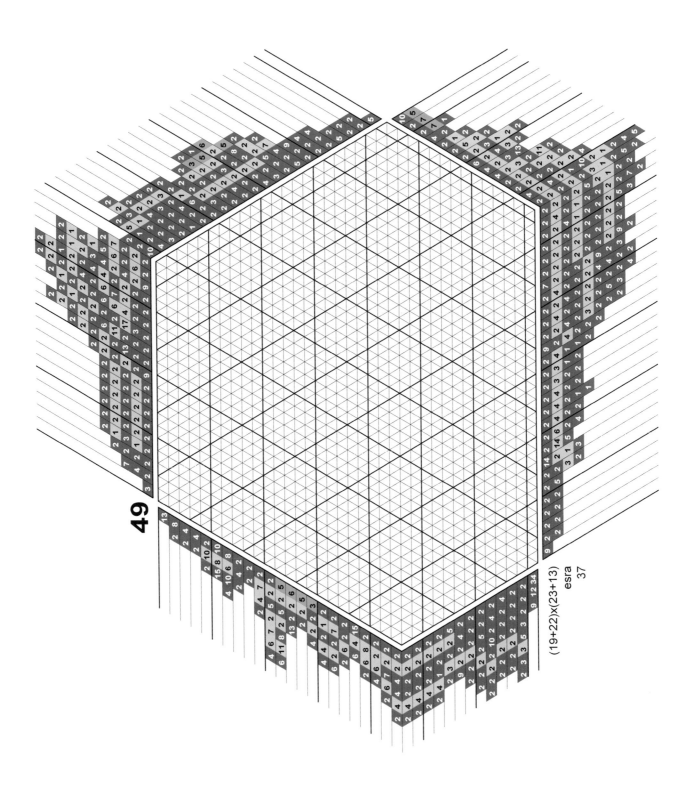

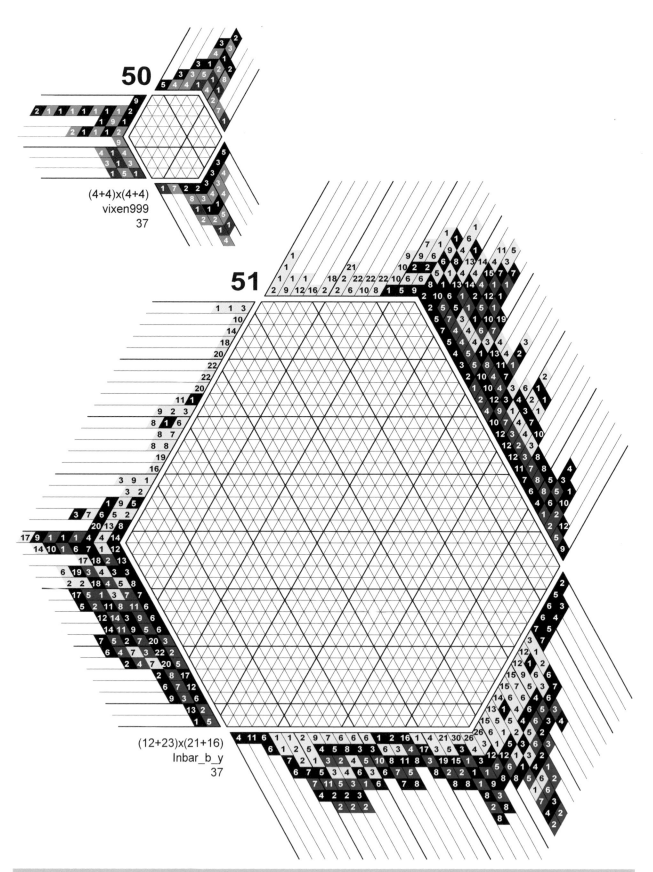

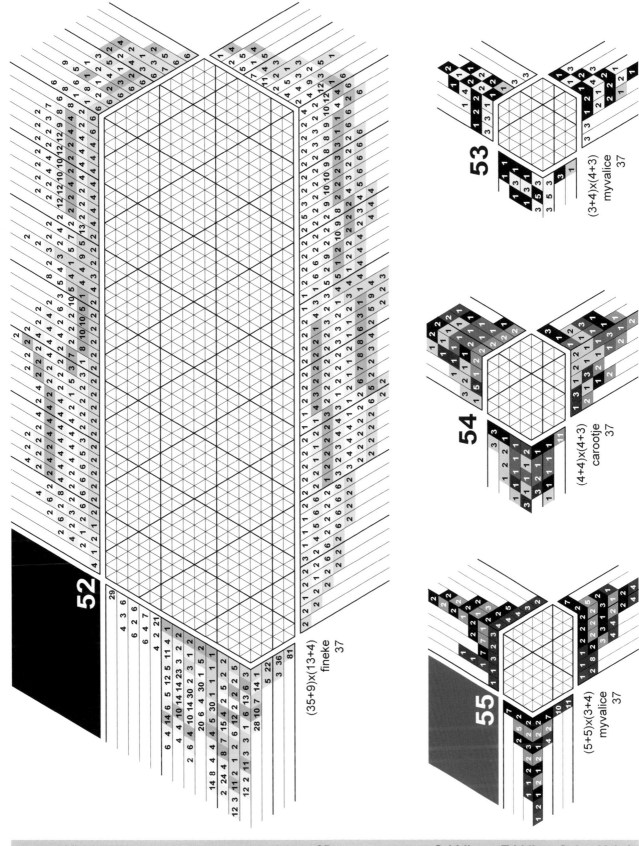

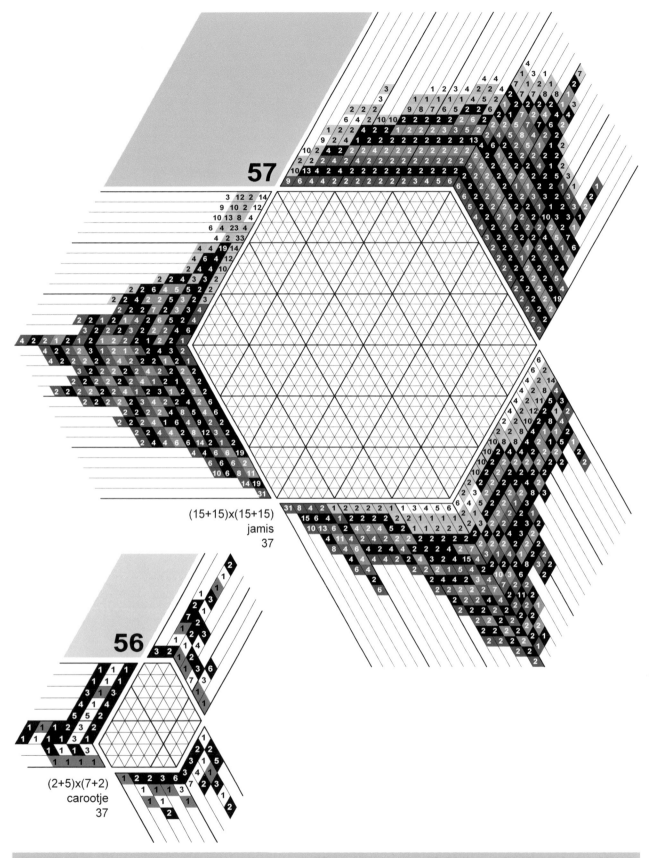

57
(15+15)x(15+15)
jamis
37

56
(2+5)x(7+2)
carootje
37

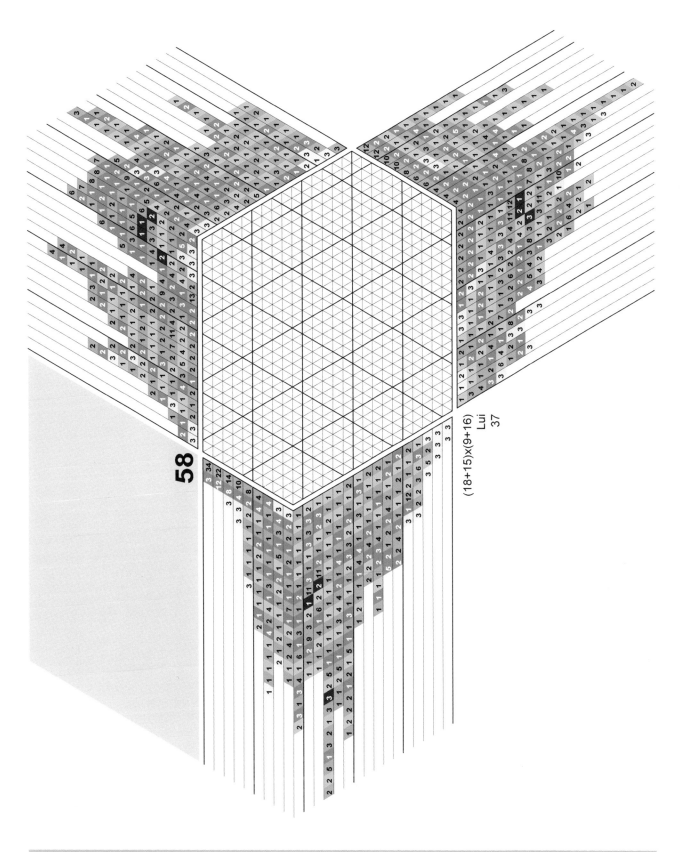

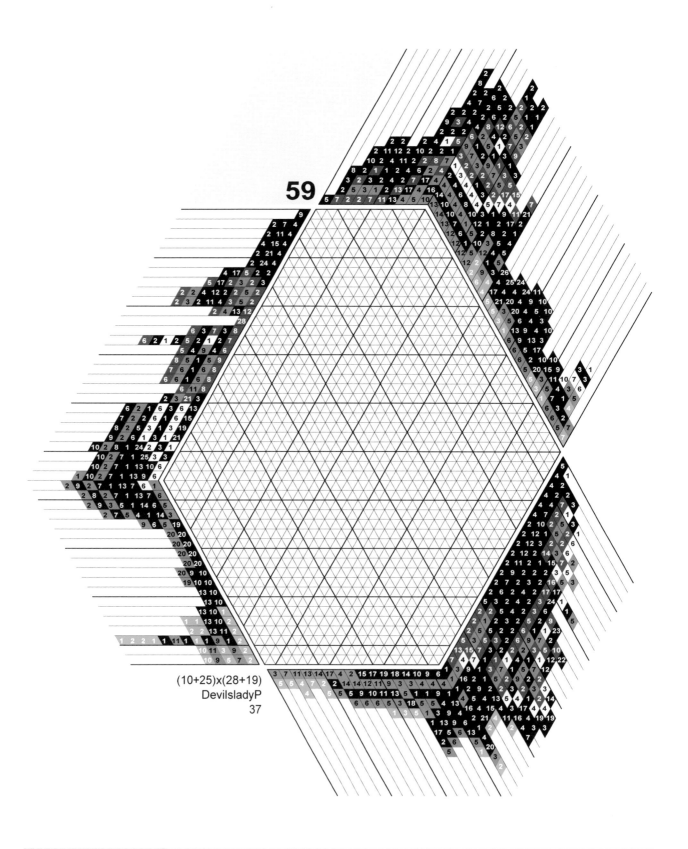

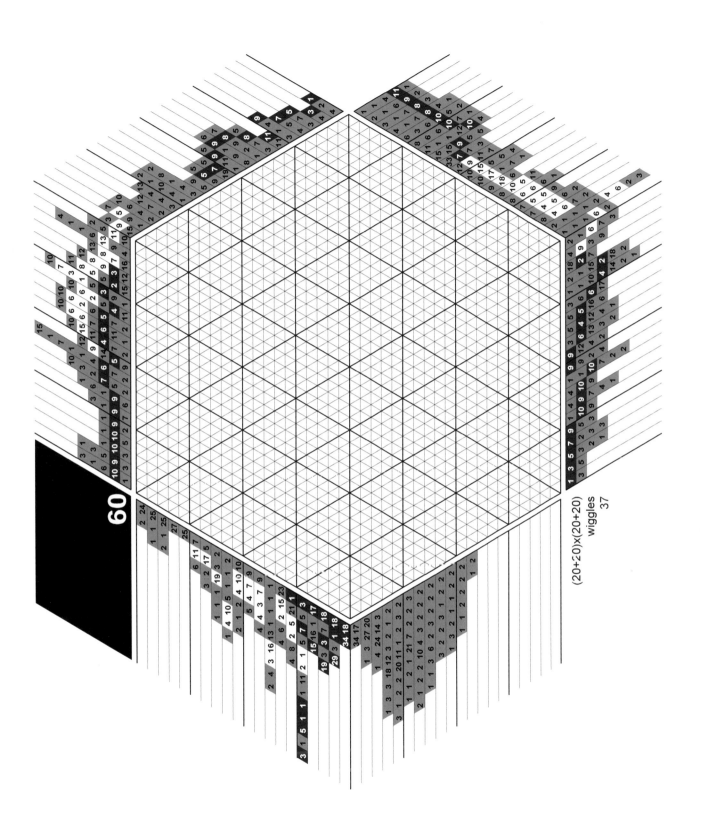

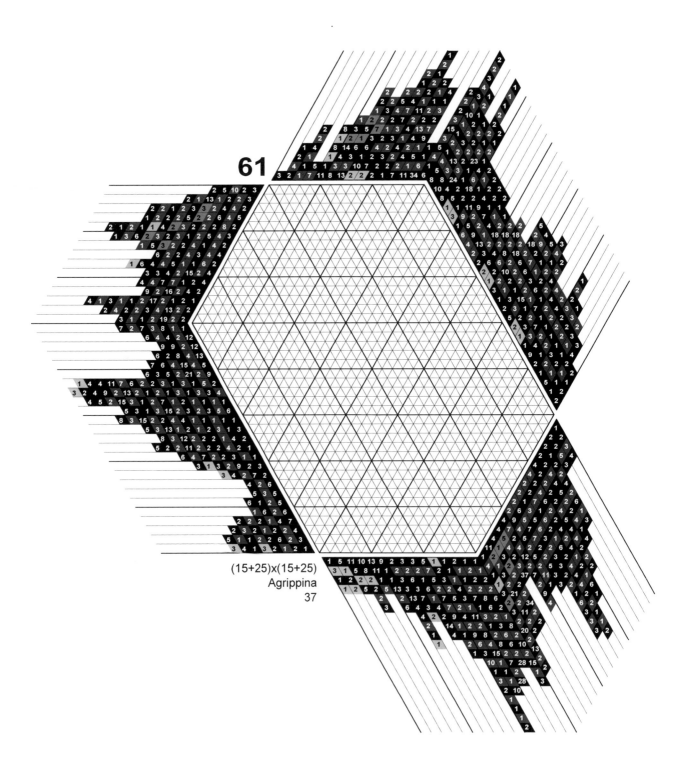

(15+25)x(15+25)
Agrippina
37

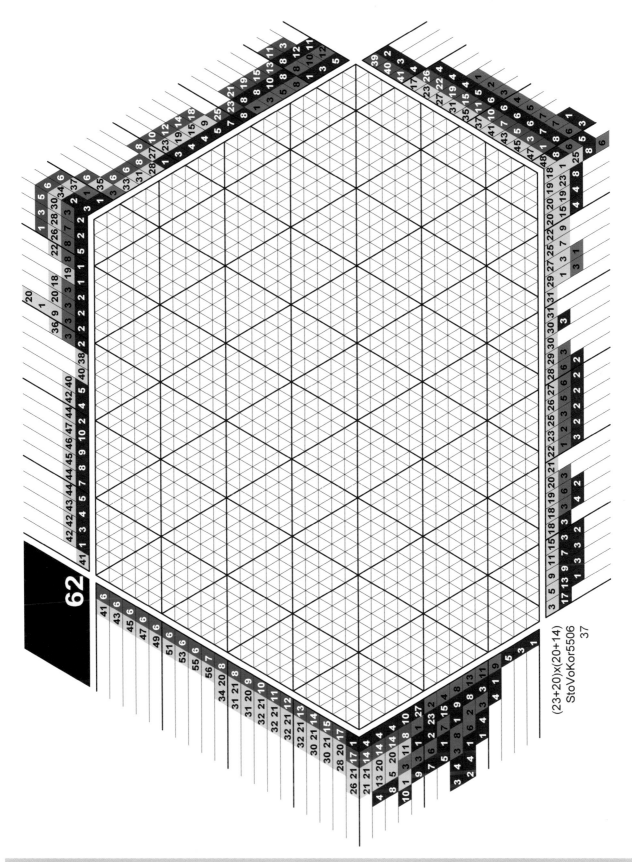

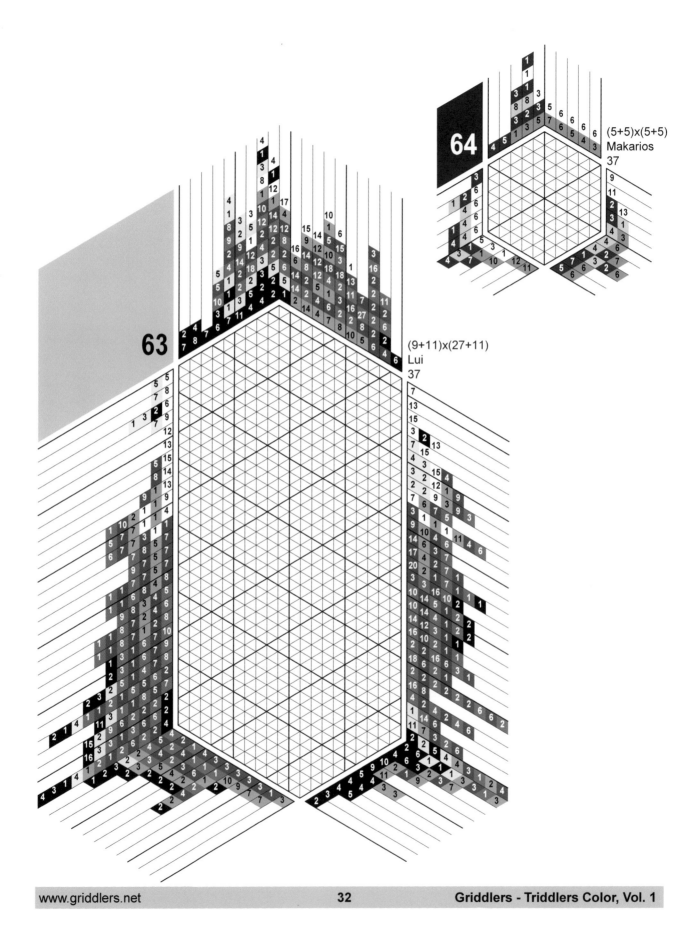

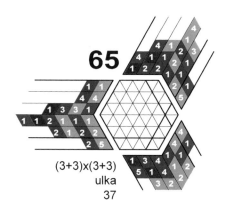

**65**

(3+3)x(3+3)
ulka
37

**66**

(2+2)x(2+2)
ulka
37

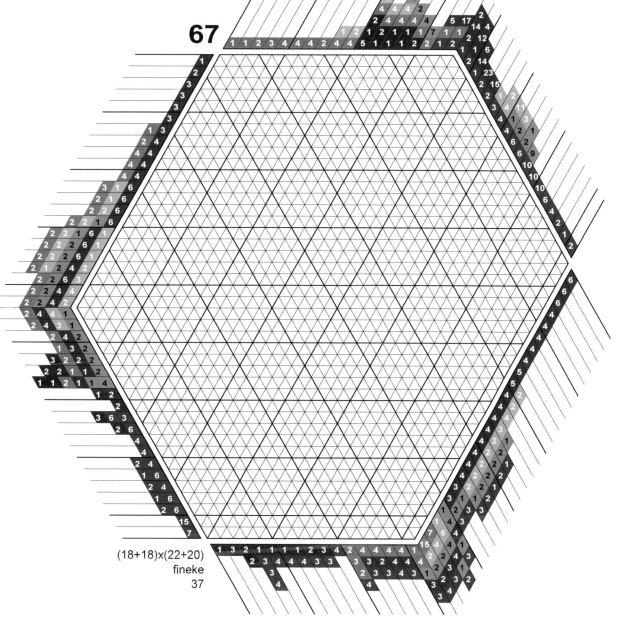

**67**

(18+18)x(22+20)
fineke
37

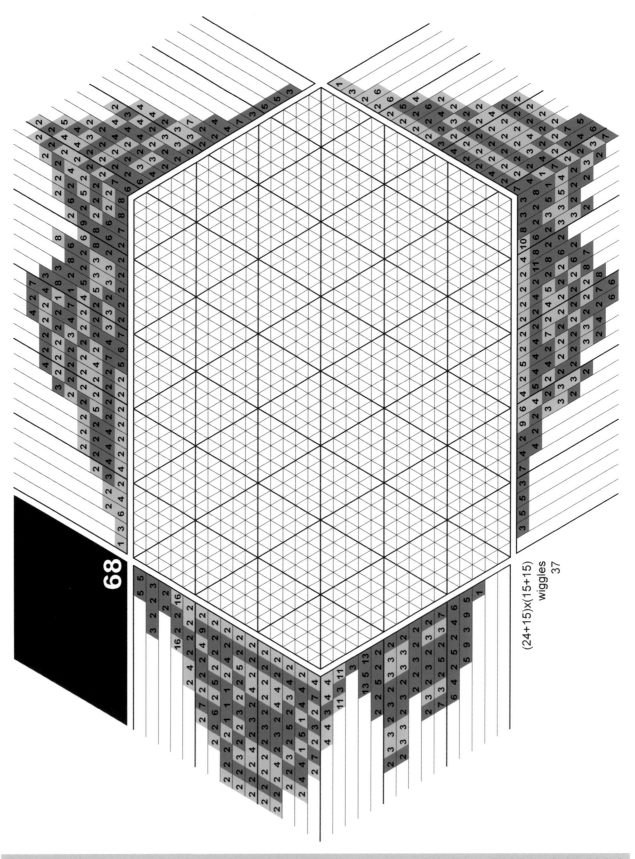

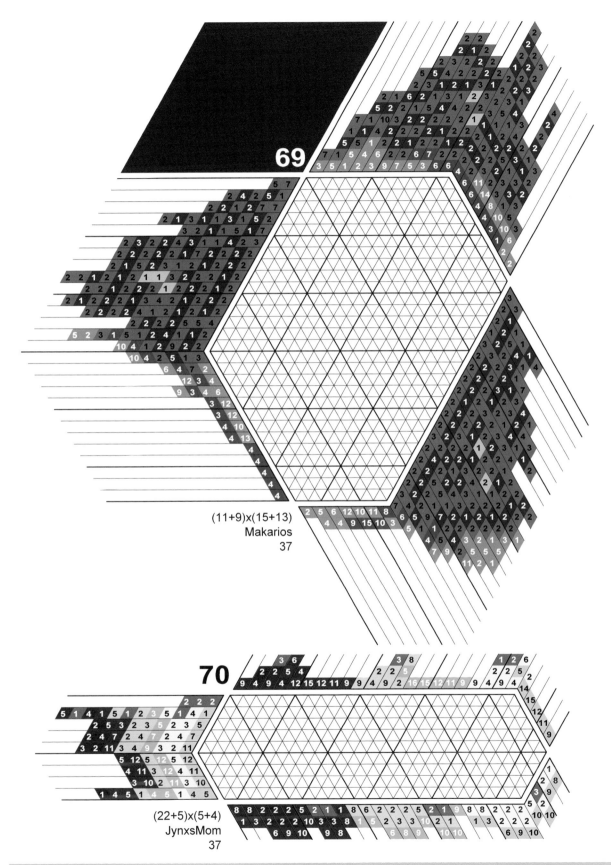

(11+9)x(15+13)
Makarios
37

(22+5)x(5+4)
JynxsMom
37

# Solutions

# griddlers
## Logic Puzzles

## Picture Logic Puzzles:

### Griddlers
**Griddlers** are picture logic puzzles in which cells in a grid have to be colored or left blank according to numbers given at the side of the grid to reveal a hidden picture.

### Triddlers
**Triddlers** are logic puzzles, similar to Griddlers, with the same basic rules of solving. In Triddlers the clues encircle the entire grid. The direction of the clues is horizontal, vertical, or diagonal.

### MultiGriddlers
**MultiGriddlers** are large puzzles that consist of several parts of common griddlers. A Multi can have 2 to 100 parts. The parts are bundled and, once completed, create a bigger picture.

## Word Search Puzzles:

### Word Search
**Word Search** is a word game that is letters of a word in a grid. The goal of the game is to find and mark all the words hidden inside the grid. The words may appear horizontally, vertically or diagonally, from top to bottom or bottom to top, from left to right or right to left. A list of the hidden words is provided.

Each puzzle has some text and underscores ( _ _ _ ) to indicate missing word(s). If the puzzle was solved successfully, the remaining letters pop up in the grid and the missing words appear in the text.

## Smart Things Begin With Griddlers.net

# Number Logic Puzzles:

### Sudoku
**Sudoku** is a logic-based, number-placement puzzle. The goal is to fill a grid with digits so that each column and each row contain the digits only once.

### Irregular Blocks (Jigsaws)
**Jigsaw** puzzle is played the same as Sudoku, except that the grid has Irregular Blocks, also known as cages.

### Killer Sudoku
The grid of the **Killer Sudoku** is covered by cages (groups of cells), marked with dotted outlines. Each cage encloses 2 or more cells. The top-left cell is labeled with a cage sum, which is the sum of all solution digits for the cells inside the cage.

### Kakuro
**Kakuro** is played on a grid of filled and barred cells, "black" and "white" respectively. The grid is divided into "entries" (lines of white cells) by the black cells. The black cells contain a slash from upper-left to lower-right and a number in one or both halves. These numbers are called "clues".

### Binary
Complete the grid with zeros (0's) and ones (1's) until there are just as many zeros and ones in every row and every column.

# Smart Things Begin With Griddlers.net

# griddlers
## Logic Puzzles

## Number Logic Puzzles:

### Greater Than / Less Than
**Greater Than** (or **Less Than**) Sudoku has no given clues (digits). Instead, there are "Greater Than" (>) or "Less Than" (<) signs between adjacent cells, which signify that the digit in one cell should be greater than or less than another.

### Futoshiki
**Futoshiki** is played on a grid that may show some digits at the start. Additionally, there are "Greater Than" (>) or "Less Than" (<) signs between adjacent cells, which signify that the digit in one cell should be greater than or less than another.

### Kalkudoku
The grid of the **Kalkudoku** is divided into heavily outlined cages (groups of cells). The numbers in the cells of each cage must produce a certain "target" number when combined using a specified mathematical operation (either addition, subtraction, multiplication or division).

### Straights
**Straights** (**Str8ts**) is played on a grid that is partially divided by black cells into compartments. Compartments must contain a straight - a set of consecutive numbers - but in any order (for example: 2-1-3-4). There can also be white clues in black cells.

### Skyscrapers
The **Skyscrapers** puzzle has numbers along the edge of the grid. Those numbers indicate the number of buildings which you would see from that direction if there was a series of skyscrapers with heights equal the entries in that row or column.

## Smart Things Begin With Griddlers.net

CPSIA information can be obtained
at www.ICGtesting.com
Printed in the USA
LVIC06n0154141217
559595LV00020BA/609